SmartMove

A Guide

NASA scientist's surprising discoveries on how to
Regain, Improve or Maintain Mobility—and Reduce
the risk of Disease Without Exercise

JOAN VERNIKOS PhD

former Director of NASA's Life Sciences

ISBN: 978-1-963565-77-5 (Paperback)
ISBN: 978-1-963565-78-2 (eBook)

Library of Congress Control Number: 2025903062

Printed in the United States of America

Published by:

QUIPPY
QUILL
info@thequippyquill.com
(302) 295-2278

Contents

Preface.................................i
Introduction............................v

The Bad Of Sitting 1

Why Worry? 2

What Is Withdrawn By Sitting? 4

How Do You Sit? 6

What Do I Need To Do To Correct This Problem? 7

Fitness Is Not Synonymous With *Health.* 12

The Solution 13

The Secret Of Gravity 14

Gravity Is Our Tuning Fork: Use It Or Lose It 14

The Case For Relaxation 16

Perpetual Motion And Staying Tuned 17

Basic Moving 18

Move! 23

Get Life Active! 24

Play 28

Take Home Message 37

Preface

Plants have figured out that they need gravity to thrive, and birds and bees to fly and navigate. Yet we humans have considered gravity as the enemy that drags us down and ages us. My 30 years of research at NASA were devoted to keeping astronauts healthy.

Astronauts' bodies go through changes as they respond and seek to adapt to living in the microgravity of space. These findings revealed unexpected clues to just how important gravity is to all of us living here on Earth.

On my research road to discovery, I first came upon the similarities between living in space and aging. Healthy astronauts lose bone and muscle 10 times faster in space, their heart gets smaller and less efficient. When astronauts return to Earth they show more age-like changes in fitness, blood pressure regulation, and balance and coordination. Healthy volunteers lying continuously in bed show the same changes as astronauts in microgravity. I realized that though we are surrounded by gravity on Earth it does no good unless we learn to use it.

Gravity is one of the four forces of the Universe and its main purpose for us is to challenge and stimulate our bodies from the time we are born, so that we can be perfectly adjusted to living on Earth. In fact, if we do not use it we will age faster. It became clear to me that understanding the significance of using gravity on earth and learning to use it effectively could put off the ravages of aging that we previously thought were inevitable (see *The G-Connection: Harness Gravity and Reverse Aging,* iUniverse, 2004).

What followed was the observation that changes similar to lying in bed continuously happen with many hours of uninterrupted sitting. This increasingly sedentary lifestyle is a consequence of the abundance of modern conveniences, appliances, and technologies. Though designed to 'make our life easier' these have instead deprived us of movement that our grandparents made throughout the day as a matter of course in each day's living (see *Sitting Kills, Moving Heals* published by Quill Driver Books, 2011). Such sitting has now been shown to be harmful to our health and to longevity.

I wrote *Sitting Kills,* to offer a different way to approach our lives and our health – one based far more on what we ourselves can do to help us feel our best.

Since it was published at the end of 2011there is new data. And people who have followed the suggestions in the book have been sharing their stories of how they have benefitted. They have discovered that they can move away from the gym without feeling guilty. They can feel better just by standing up once in a while at the office, experiencing immediate benefits. This has been informative and very rewarding. Since *Siting Kills* there has been lots more data analyzing the relationship of sitting to death and illness, but not much in explaining why sitting is bad for you and how to fix the problem.

Time and again I heard the question 'But What Do I Do?' This concise companion Guide to *Sitting Kills* is designed to answer this question you raised. If you do sit a lot and move less you are at huge risk of illness and incapacitation. More importantly, I wanted to summarize the solutions that are available to you right now without expense or going to the gym. It boils down to moving, yet moving smartly – making the most of the movement needed to accomplish whatever the demands of each day. Moving smartly means understanding why and how each move works to add a health benefit to you. Moving smartly also means moving with purpose, moving to get something done that needs doing. Such movements have the advantage of fitting into your daily schedule wherever you are. They are flexible and convenient, requiring no special equipment, place, or schedule so that you do not have to set aside time, change clothes, or shower. What's more, they give you a huge sense of accomplishment by getting your jobs done as well as benefitting yourself in the process. I have condensed them down to five basic types with suggestions for variations that suit you, your life, and the occasion best.

Introduction

I was lucky enough growing up in the 40s to be surrounded by people who as I look back, seemed to move all day long. We took the tram to school twice a day and had to walk the last bit to get to the school. We went home for lunch and back to school for more classes in the afternoon. We played some sports every day, field hockey, track, swimming, netball (a girl's version of basketball), and some played tennis. After hours my sister and I trained and swam competitively at a local swim, rowing, and sailing club. It depended on the season. We had an elevator but usually raced further up the two flights of stairs and never took it down. There were no TVs, nor computers, and no cell phones. Radio was our entertainment. And yes, whenever the teacher came into class or left at the end we all stood up in deference. So we did that 16 times a day without giving it a second thought. Oh yes, and we were all scored on posture.

My contemporaries are living longer. Modern medicine is keeping us alive but the quality of life is not what might be desired. However, people who don't move, die sooner. They suffer from illness more and weakness that becomes incapacitating. I would like to be healthy and active till I drop dead. Wouldn't you? That is why I am writing this Guide. I want you to feel great now and live longer and better. In *Sitting Kills* I have given you more than enough of the supportive medical explanation. In this Guide, you will get the tools to enable you to respond to the demands of daily living without even exercising.

Why are we sitting more? Modern conveniences have made it easier to get things done for us around the house and garden. Your car transports you to work. A machine washes and dries your clothes. Work for most of us is sitting in front of a computer all day or involves driving a long commute in a car or even driving next door to a shop. For entertainment, we stay at home watching TV, controlled from our chair by a remote, with pizza for dinner delivered to the door. We therefore do not need to move the way we used to. In fact, modern conveniences have robbed us of the movement we spontaneously did in the act of living.

Why move? Because you intuitively know that getting to move more would improve your health, change your life, and even save your life. You will enjoy greater health. What's more, you will feel good. You cannot feel good if you are not healthy and it's hard to feel happy if you do not feel good. Sixty-year-old Elaine who followed my advice said "I feel 20 years younger. I have walked every morning but was not aware how much I sat the rest of the day. Now I am moving around all day long and still enjoy my TV."

Sitting is OK as long as it is interrupted often. New data shows uninterrupted hours of sitting increase the risk of cancer, diabetes, and cardiovascular disease even if you exercise regularly. Between TV and other media, we are encouraged to sit for many hours. "Jack sits in front of a computer screen all day long. He does not care to exercise because he has back pain but once he understands how that works, he can go along with the idea of standing up every so often and stretching up."

It's not about something being 'good for you' but, something being better than nothing. Whatever your starting point, it's something you can do right away, that makes you feel and look better, that helps you feel alive. Nor will your muscles and joints feel sore.

You may work in an office, be unemployed, drive kids back and forth, care for parents, or are recovering from injury. Maybe you have already noticed changes in what you are able to do or cannot do and you want to be strong enough to carry out everyday tasks again.

The solution is easy and readily available. My research at NASA in keeping astronauts healthy uncovered the unsuspected medical connections between the health hazards of living in microgravity in space and the chronic diseases caused by sedentary lifestyles here on Earth. The most revealing finding was that movement that resists gravity is essential to good health. Thinking of gravity like an astronaut, gets you moving. All-day activities involve moving in a way you would normally do in the course of living. Many of these movements like dancing, gardening, walking with your friends, walking to the shops, lifting a child, rolling out dough or baking cookies, swinging on a swing, or jumping on a trampoline, are also fun. They all count.

The secret is interrupting long stretches of sitting and turning these movements into habits. Habits take eight weeks to stick and they are free or low cost. Variety and purposeful motion take away the boredom often associated with exercise.

Now let's get down to basics: We were designed to move. We have forgotten that "the gift of life is motion." Our body and mind are a perpetual motion machine. We breathe, our gut pushes food around, our heart pumps, thoughts race, eyes move, nose sniffs, face smiles or frowns, mouth chews or talks. We are designed with a spine that swivels along the gravity vector and arms and legs that bend and reach. If we don't move any little part of us, it rusts, jams, or quits. Forcing it to move, hurts. We all know how pain feels. So what do you think is the answer?

There are simple easy ways to cut through complicated exercise routines that require equipment or personal trainers. Digest the idea that you are a machine; you are a machine of moving parts.

Think of moving, all day, every day, doing things around the house, in the garden, bending down to pick up something, shoveling snow, stirring a pot of Sofia Loren's favorite pasta sauce, Reaching up to get something off a shelf. Hang up clothes to dry on a line if you can. Take a class of yoga to stretch every which way and learn to breathe better; fill those neglected lungs and stimulate those neglected rib-cage muscles and diaphragm that seem to work in spite of us. Discover how much better they would do their job if you used them consciously.

Moving in the course of living keeps your body tuned and oiled ready to respond when you need it. Simple everyday moving is the foundation of health. It's what keeps you alive and healthy. It makes you Feel Good!

If your mobility is constrained you can experience improvements just by standing up every 20 minutes. Even if you cannot stand up at first you will regain your mobility in a few months if you keep trying, or even if you keep thinking 'Stand up.' If you cannot get up at all there are many stretching and strengthening activities you can do while sitting. "I find that sitting on a straight-backed chair instead of a comfy chair has improved my posture and strengthens my core muscles."

Studies show that too much sitting is killing you even if you exercise. Why isn't exercise enough? What is missing? Uninterrupted sitting means that you have taken away the frequent stimulus you need every so often. The stimulus is provided by the change in posture, your body working up in the opposite direction to the downward pull of the force of gravity. How often? As often as you can spread it out throughout the day. What hours of sitting take away are the crucial interruptions that challenging gravity by standing up often provide? Even those who exercise three to five times a week but don't move the rest of the time, are at risk and can benefit from this kind of moving.

Each one of us is different, has different needs, and a different perspective of what feeling good means. But you will find something in this Guide for everybody. It is not like other guides a set of exercises you need to do every day with many repetitions. Here is where you begin. You were designed to be a perpetual motion machine, to move all day, every day, including weekends; to move constantly as a matter of course without setting aside a chunk of your day to go to the gym, change, shower, or drive. Instead, you can move at any time as part of your daily tasks, tuning your body as you move about, like tuning your guitar. Set it aside for a while and it needs tuning again.

Where do you begin? This guide is about building your foundation of movements that challenge gravity, plain and simple. You can do this. Unless you are in a coma anyone can move, however small or large that movement is. More importantly, it is designed to be built into your life and your habits. It's something you can do right now to reduce inflammation caused by hours of uninterrupted sitting. Moving makes you feel better right away. You'll have more energy, stand or sit up tall, look and feel younger, and surprise yourself with the compliments you get.

There are a few easy steps you can take to raise your sensitivity to how you feel and what you do:

Jot down how long you sit every day.

How many of those hours do you sit continuously?

Now that you have become conscious of your sitting habits, consider what are your moving habits. What do you do and when do you do it?

Next, make an effort to interrupt your sitting and reduce your total time of not moving. Then refer to the Basic Movements section of this guide for suggestions on how to incorporate them in your day.

Increasing movement in small ways yields big results. This Guide gives you a stand-alone approach that is part of your day as you go about the business of living. People from 30 to 70 who don't have time to take care of themselves, who are becoming aware of the risks of too much sitting, who have stopped moving on a daily basis as new appliances and devices have made life easier, will rediscover the benefits of moving – no injuries, no equipment, no time to have to find to take care of yourself, little effort at no cost. A handy side benefit is more free time, more money in your pocket, staying trim, and feeling energetic all day.

Whatever the reason you decided to do something about your health, you made a good decision. Becoming more tuned in to yourself, your life, how you spend your time, and what you do is the first step to a stronger healthier you. You will find more technical information in my book *Sitting Kills, Moving Heals*. But **this** *Guide* tells you what to do and why and how so that it can become second nature. Your life will change with new habits and attitudes. Believe me, you will not only notice the dust on the table but you will spontaneously take care of it right there and then. Done! Your sense of accomplishment and control will grow with each little step.

Health is not the same as Fitness. Sitting all day even if you exercise takes away the benefit of exercise. Bear in mind that health is different from fitness. Health is how you feel and function. Fit is what you are able to do. Exercise makes you fit. Moving keeps you healthy. So moving is the foundation; you cannot live without it. But you can and should build on it with exercise whenever you can. They work in different ways and complement each other. If you only do what is in this Guide you may need to add some exercise to gain endurance or strength. But it does not work the other way around. If you only go to the gym for exercise you may not necessarily be healthy and would need to build your foundation of all–day moving for good health as well. This Guide is about the basics that keep you healthy and resilient. This Guide is the answer to your question: "What should I do?" and "What should I do first?"

Where Are We Today?

We Live Longer But Not Better.

Americans have struggled in recent decades to exercise more and eat less, but one thing hasn't changed: we spend hour after hour each day virtually immobile in our chairs or cars and we're heavier, sicker, and more tired than ever before. The way we live is killing us, and we can't seem to stop it.

Sitting is now considered an independent risk factor for many chronic diseases and premature death. This means that on its own even if you do not smoke, are a heavy drinker, and even if you exercise regularly you are at serious risk of illness and early death if you sit continuously for hours on end.

We live longer than ever before, but we are not living healthier. On average we live 25 years longer than we did 100 years ago. The average lifespan was 55 and now is 82.

While death from many diseases including heart disease, cancer, diabetes, and others has declined, the incidence of these has increased overall. There are more and more people living with the disease now than there used to be. The number of adults with these diseases has increased from 9 to 30% in just the last 10 years (since 2003).

On average we are fatter. In 20 to 74-year-olds, the rate of obesity rose from 25 to 35% between 1980 and 2010 alone and is far worse now. Obesity is a major risk factor for all chronic diseases.

More of us are experiencing neck and back pain. Statistics would tell you that about half of you have neck or back pain at some time.

Our well-being is suffering too: The number of Americans reporting challenges with depression and /or anxiety has steadily increased over the past half-century.

So What's Going On and What Can We Do About It?

Modern Sedentary Lifestyles Undermine Our Health

We move less today than at any time in human history.

Today's comforts, conveniences, and forms of electronic entertainment deprive us of the simple everyday movement of our recent ancestors. The technologies of modern times have produced appliances, tools, and work environments that we are told make our lives easier. They have reduced our level of physical work and so we sit more than we used to and more than we were designed to sit. Since the

Industrial Revolution, the work it took to do common household chores or to move about has been minimized by wonderful modern conveniences in the home and at the office. The dishwasher, the washing machine, the vacuum cleaner, the food processor, and cars – they 'make our lives easier' but rob us of frequent daily movement.

Americans today spend more than half the waking day sitting – more than ever before. What has changed?

Americans now spend an average of 13 hours/week on the internet, while TV time has increased to 34 hours/week. This is despite nearly half of us saying "We watch too much TV."

In the office, we've almost completely stopped moving. Until the 1990s working in an office meant walking from office to office to speak with people. Today electronic communication, your cell phone, has replaced most of this.

We sit during commutes. Until 2000, time spent commuting had consistently increased since mechanized commuting began with the Industrial Revolution in the mid-1800s. Before that people walked to work and children walked to school.

The Bad of Sitting

Much of the increase in chronic disease in recent years is now traced by research to excessive sitting. A recent study spells it out: 200,000 Australians aged 45+ were observed over a period of time. In the three years of the study around 5000 of those people died. About 7% of the deaths could be attributed to prolonged sitting. Those who sat for more than 11 hours a day had a 40% increase in risk of death compared to people who sat for fewer than four hours. Those who sat for 8 hours or more/day had a 15% increase in their risk of death. And this 15% accounted for other risk factors too, like age, smoking, and more.

The more we sit the more tired we seem to feel, and the less energy we have. This is particularly evident in the office where sitting is often uninterrupted for most of the day while we are connected to various wired and wireless devices. Then we go home in the evening where TV and other home entertainment perpetuates uninterrupted sitting. Health and productivity are affected by more sick days, increasing our physical and emotional discomfort as well as the cost of keeping ourselves well.

The fantastic news is that it's possible to reverse these health impacts by being more like our ancestors: by using simple, frequent movements that challenge gravity.

How many hours a day do you sit?

People who spend two or more hours sitting in a 16-hour day are more likely to have health issues. Imagine what happens when you sit 10 to 12 hours a day. Most of us spend as much as 55% of our waking hours each day sitting – in our chairs, driving our cars, working at a desk at the office or at school, not to mention sitting on the couch watching TV.

Why worry?

Studies show that the more we sit, the greater the chances of dying of heart disease, cancer, stroke, or diabetes. You can get back pain and joint pain at the office. We are fatter, sicker, and more tired and unhealthy than ever. Children who learn to spend too much time sitting are bored, restless, and depressed, and show up with diabetes, heart conditions, obesity, and other signs of poor health earlier than we used to see in people over 50.

The problem is compounded by:

- Beliefs that a pill or the next medical breakthrough will make us well.
- Long hours of sitting. – in the workplace, remaining electronically connected to those around us. Outside office hours extended sitting whether at home or in vehicles.
- A belief that regular exercise counteracts the impact of uninterrupted sitting. Research now shows that this is not the case.
- Many people don't or cannot keep up a regular exercise program, classes or even walking.
- Studies show that more than 2 hours of continuous uninterrupted sitting, even when combined with regular exercise, are linked with death from breast or colon cancer, increased incidence of diabetes and obesity, heart attacks, and stroke[1].
- "Avoid prolonged sitting, too," said Dr. Freddie H. Fu, a professor and chair of orthopedic surgery at the University of Pittsburgh Schools of the Health Sciences, "since uninterrupted

sitting leads to stiffness that aggravates the pain of going downstairs."

- Falls are a sure way to an early death. One of the best ways to prevent falls is to maintain a sound, responsive, and enduring sense of balance. Good balance starts with a good gravity-aligned posture whether standing or sitting.

So you are telling me to exercise?

Why is it that sitting down most of the day, despite a strenuous morning workout, can be as bad or worse than smoking?

By all means, if you are exercising, please keep on doing it. Yet although about 52% of Americans claim they exercise for at least 30 minutes three or more times per week, they still sit the rest of the time as much as those who do not exercise.

Tip: *Prolonged sitting does something that is distinct from not exercising.*

What is withdrawn by sitting?

- "Sitting too much is not the same as exercising too little," says Marc Hamilton at the

Pennington Biomedical Research Center in Baton Rouge, LA.[2]

- One hour of vigorous exercise daily, does not compensate for the effects of sitting the rest of the day on insulin and blood lipids that are risk factors for diabetes and heart conditions.[2]

- A 20-minute bout of yoga stimulated memory and brain function better than 20 minutes of running on a treadmill[3]. The practice of Yoga is about moving *naturally* not exercising.

- Nor is standing for as many hours the cure for sitting. People who need to stand on their jobs suffer from aches and pains, swollen ankles, and blood clots in their legs.

My research showed that

Tip: *It is not the number of hours sat that is the problem but the number of <u>continuous uninterrupted hours of sitting</u> that are the cause of modern poor health.*

The Alexander Technique is a method of learning <u>how you move</u> when you do anything that involves "moving" (i.e. walking, running, sitting, lifting, eating, knitting, golfing, swimming, sweeping, and on and on…). You learn where you are over-tensing or over-contracting your muscles and therefore your joints and creating stiffness and pain. However, we build habits over our lifetimes, and even though we have been "over tensing" (I think of the steering wheel, shoveling snow, the broom, or the coffee cup!); we may not know it because it has become habitual and not in our sensory awareness. And now we have stiffness, pain, and sometimes arthritis in our wrists, fingers, and spine and we don't relate it back to our "over-tensing" of our muscles and our joints.

When we learn the Alexander Technique principles of how to bring our habits into our sensory awareness and change them; we create much more ease in our arms, wrists, spine, hips, neck and shoulders, etc. We learn we can drive without over-gripping the steering wheel, pick up a cup without grabbing it with a death grip, and still get the task accomplished! And we keep arthritis and chronic pain away! (Tensegrity and mechanotransduction. *http://karenloving.com/tensegrity-is-our-model-to-find-resilience-in-our-movement-and-a-spring-in-every-step/*)

How do you sit?

It is not just the hours of sitting that are bad but how one sits. Comfortable furniture and recliners encourage greater curvature of the spine. Supported office chairs discourage sitting up straight. Shoulders and back hips are all vulnerable points. Stooped head while sitting or even while walking is detrimental to overall health and frequently a source of back and neck pain as well as flabby core muscles.

What do I need to do to correct this problem?

It is not exercise at all. One hour of daily vigorous exercise cannot compensate for the effects of sitting the rest of the day.[2]

Just plain MOVE. Moving is the host of non-exercise activities that we do all day, every day in order to live and function. We need to move to be able to get about or run away from danger when we need it, or when we jump with joy; to cook and play with your child or grandchild on the floor. They are movements like standing up or sitting up, that are required to keep our body tuned.

Staying Tuned. The value of these basic movements is to keep our body tuned. Stay sitting all day and your body, like your car or your piano, gets out of tune – out of synchrony. And when you suddenly need to stand up or react you find that you are unable to do it.

Staying aligned. Body alignment is basic to living in gravity. We are designed with hinges that need to be oiled and used or stiffness and pain result. The same happens if you twist and strain them. Stop and think about all the ways your body is designed to articulate; then think of what it would be like if you could not make that movement. We have created tools and ways to eliminate using our joints correctly -- chairs, beds, desks. Look at what is happening in Japan where advancement means a bed, a couch, a chair or a table, instead of a *tatami* or kneeling on the floor to talk or sip tea. Think of all the movement that is eliminated.

Gravity is just as important when sitting as when standing. Sitting with the head squarely supported by the neck and body, looking straight ahead, periodically sliding elbows back or stretching the arms to reach the ceiling are all relieving moves to release tension, maintain good sitting posture and tone core muscles.

Moving Mindfully. Moving with a purpose is more likely to be sustained. Hike up a hill, climb a tree, walk to school, cycle to work, get outside and into nature more often. Be social, hold someone's hand, and do things with other people. By moving mindfully, it is more likely you will continue to remain active and not get bored and stop.

Tip: *Involve your brain and your emotions. Benefits come from moving with purpose, in a context that involves our senses and intellect. Movement is best when we interact* *with the environment, with other people, other living things, that we go somewhere or do something in the pursuit of living. Healing Movement is all about the <u>life activities</u> we do – because we enjoy, want to or need to do them.*

<u>MFH is all the small, brief yet frequent, low-intensity movements that we would not consider as exercise and that one makes throughout the day naturally while we go about living.</u> Standing up is the most effective of these. Raising a mug of coffee, reaching for something, picking up a child or kneeling down to pray, are not what we would consider exercise. James Levine at the Mayo who named this type of movement NEAT (Non-Exercise Thermogenic Activity)[3] found that *lipoprotein lipase*, an enzyme that transports fat into the blood for clearing, is seriously reduced during prolonged sitting. And that these small movements, primarily standing up, caused this enzyme to increase, thus accelerating the metabolism of fat.

But Levine was stuck on the traditional burning of calories. The T in Levine's NEAT stands for thermogenesis (heat production) – whenever we move around just as when we exercise, the calories we consume are converted into energy by contracting muscles and generating heat in the process. Thus, people who move around a lot all day even if they don't go to the gym or exercise intensely, may burn up as

9

many calories in the course of a day as people who sit all day. The difference between calories burnt by exercise and those by non-exercise MFH is that with exercise the calories are burnt during one intense period of activity rather than throughout the day. Another difference is in the fuel used – exercise uses mostly sugar, MFH mostly fat.

With MFH, calorie burning is not the purpose but a bonus.

Exercise and Moving for Health differ in four significant ways:

- o The effectiveness of MFH relies primarily on using gravity.
- o It requires cognitive, emotional, and physical interaction
- o It requires a purpose – a mindful interaction with nature, the immediate environment, other people, or pets. Turning a job into fun.
- o It is practiced throughout the day, every day, throughout life.

In my own research, we found that

- In fact, standing up was more beneficial than walking for blood pressure regulation and stamina.
- It was not how long one stood up but how many times – at a minimum about 16 times a day from lying down, or 32 to 36 times per day – from sitting.
- Standing up every 10 or 20 minutes throughout the day seems to be the most effective.
- The signal was the change in posture triggering the inner ear brain-balance system – the vestibular system – and blood pressure sensors in the heart and neck. Ideally, these signals should be throughout the day, every day, including weekends.
- For maximum benefit think Change, think Posture, think Gravity.

The force of gravity pulls straight down to the center of the Earth. We were designed to live in gravity and use it to stimulate every part of the body and maintain it *tuned*. We are most relaxed when aligned with the gravity vector. A baby is born into gravity and develops to function best by using the gravity vector. Once fully developed, gravity is used to maintain all systems responsive. It is this *tuned* foundation on which exercise as we know it, can build strength and endurance.

Fitness is not synonymous with *Health.*

You can be healthy and dynamic without being fit to run a marathon or do 20 push-ups and you can be very fit, even a top athlete, without being healthy. Without a solid well-tuned foundation, the benefits of exercise cannot be sustained, you can rapidly decondition when you stop exercising and injuries can readily result. MFH works in different ways than Exercise. Exercise uses intense muscle contractions to burn glucose and generate energy for endurance. The movement for health relies on mostly low-intensity sustained activity that uses the gravity vector for body tuning, balance as well as fun, acceleration, and exhilaration. By recruiting the brain and the vestibular system in a way that exercise does not, it integrates the sense of direction, acceleration, cognitive, visual, proprioceptive, touch, and pressure elements into this total body grounding.

That's what using Gravity movements builds on and reinforces for a total you. The sooner and better you can adopt these movements as Gravity Habits the healthier you will be and the better you will feel. Ignore gravity and it will drag you down and age you.

The Solution

<u>Enrich your life with Healing Movement</u>.
By introducing these daily and all-day into your life with a focus on using **gravity** to your advantage you will rapidly feel better and full of energy. Most importantly you will naturally and cost-effectively maintain or regain your health and independence.

- Frequent all-day movement that challenges gravity is more essential to good health than traditional sports-like exercise.
- <u>*Standing up, and changing posture*</u>, is the simplest, most effective gravity-using activity. Gravity is the tuning fork that stimulates every part of the body to keep it vibrant.
- <u>*Move*</u>. This is simply about any movement.
- <u>*Life activities*</u>, in contrast to sports-like exercise, include small seemingly inconsequential everyday movements that humans traditionally use to accomplish the tasks of living.
- <u>*Play*</u>. Gravity is at the core of all our fun activities.

Tip: By once again <u>making these movements habitual</u> while consciously adding others, strength and balance, mobility and a sense of well-being are the result.

The Secret of Gravity

The way we look, feel and function on Earth is due to gravity. Ever since we are born we develop and grow to remain best adapted to live in Gravity. The price we pay for not floating off into the universe is gravity that keeps us grounded on Earth. The price we pay for being healthy on Earth is that we constantly have to move in gravity and use it to remain healthy and mobile so that we are not rooted like plants but can move around and be independent as long as we live.

Gravity is our tuning fork: Use it or lose it

Think of gravity as this virtual rod pulling us to the center of the Earth. It does not change. If we let it have its way we would be crumpled on the floor. The fact that we are standing or sitting means that we are moving and contracting muscles and bones that support us upright, even though we are not aware that we are doing so. It is how we move and rotate around this force that determines whether we benefit by using Gravity less or more. We say that the pull of gravity on Earth has a force of 1G. We can increase the magnitude of the effect of this force by jumping for instance. From the moment we are born, we move to develop maps that are etched in our brains and tell us where we are relative to our environment. These become weak and erased with time if not reinforced.

Using gravity as the stimulus always in relation to our environment keeps our brain and body tuned.

<u>Gravity gives us.</u>

- Our point of reference – up or down or where we are relative to our surroundings
- Our downward stimulus if we choose to challenge it
- Our sense of weight and load
- Our sense of direction
- Our sense of acceleration and fun
- Our ability to relax totally
- It centers and grounds us

What kind of Plan should I follow?

<u>The Doable Plan</u>. One that incorporates Movement but that does not have to be a certain way or a certain number of steps. A plan that is your own is more manageable. The more individualized your plan, designed to fit your life and lifestyle the more likely it is that you will do it. If you feel good as a result of what you do with a sense of accomplishment, the more likely you will be doing the things you need to do anyway. You will enjoy doing things you used to consider chores which will end up having a positive effect in your life. And you can use that extra time freed up from exercising, showering, changing, driving to the gym, and the extra cash in your pocket, on your family, your friends or on yourself – pampering yourself with a massage or a few minutes of meditation.

The Case for Relaxation

When we are sitting are we relaxed? Not necessarily. We can only be relaxed when we give in to gravity totally. If we stand *still*, sit or lie in bed to sleep or even during sleep, some part of us is always moving and contracting. Therefore, any muscle contraction begins from an already contracted state. The nearest we come to total relaxation is when we lie on the ground, completely give in to gravity, and quieten our brains. We can also feel total relaxation sometimes when we float in water or are in space without gravity.

Free of Gravity

My friend astronaut and cardiologist Dr. Drew Gaffney told me that when he was launched into space in 1993 on the SLS-2 mission, as soon as the Shuttle separated from its boosters, he felt an immediate sensation of total freedom like he had never felt before or since on Earth, even though he was tightly strapped down to his seat

---

Perpetual Motion and Staying Tuned

Like standing up and sitting down or anything that includes a change in posture or varies our relationship to gravity, which is essentially any movement, tunes us. The less we move the less we use gravity.

For people who make the conscious choice to move more, regular use of the following activities in various combinations will have positive impacts on your health.

Add these activities to your current daily routine. Each of you will find ways to use some or all of them. Generally, the more variety the better, but start with one or two at a time and keep adding. This is intended to be fun.

Revel in the benefits of making small and fun changes to your day. The body needs perpetual motion to keep us stimulated. Motion that challenges gravity, until they become a habit, like brushing your teeth is a habit. You alone have the power -- Use it!

BASIC MOVING

These movements are excursions up and down in the direction of the <u>gravity</u> vector. They can range from squatting or sitting to standing up to jumping, challenging gravity all the way. Make use of the downward motion as well, sitting or squatting or even sitting on the ground but moving slowly and resisting gravity all the way down to tone your posture muscles—the muscles that support you. You will notice that these are all movements you can and should do repetitively throughout the day. Use them like your body's tuning fork.

1. Move in the line of Gravity

i) Stand Up

This is about standing up without going anywhere. Change in posture is the most crucial signal to body health and the single most important activity you can do.

Let's face it, most of us spend a large part of our workday sitting. Our bodies seem to adapt to sitting for long periods of time, and it's easy to stay glued to our chairs until we go home. But don't! Take the time to get up from your chair periodically. Not only does it invigorate your body and help you feel good, but it also helps it release muscle tension and gives you an extra boost of energy that can result in better concentration and more high-quality work. Judi Bar, Yoga Therapist

o Standup every 20 minutes during work for 1-2 minutes. This could include walking to the restroom getting a drink of water or picking up a fax or something from your printer. Standing up without walking is all you need.

o If you work in an office you may need to be reminded when it is time to stand up. The Firefox browser Add-on "Take A Break" will alert you with a flashing icon or a pop-up window at intervals you define when it's break time. addons.mozilla.org/.../take-a-break.

o It does not matter how you get up to begin with, whether you lean on something, the chair, your knees, or not. The aim is eventually to be able to stand up without using your arms to push off, arms hanging by your side and moving straight up.

o The easiest way to start is to use a hard-backed chair.

o With practice when you can fully control your motion unaided, try getting up in the same way from a couch or an armchair, still without using your hands.

o Practice this wherever you are.

o Try standing up at your desk when you receive a phone call or during a meeting.

o Structure your workspace or sitting area to introduce necessary occasions to stand up (e.g. place your printer out of reach, ditch your TV's remote control, or stand up during TV commercials, etc.)

o When you go to the grocery store or mall, park in a space farthest away.

o If watching TV, get up to have a glass of water, cook, or wash the dishes during commercials.

o Dr. Mercola goes one better – he finishes off his standing up with a jump straight up. This generates up to 6G as opposed to the 1G of standing. You don't need to do this every time you stand but it adds variety.

ii) Sit Down

Each time you sit down do it mindfully, as slowly as you possibly can. If you give in to gravity you will just plop into your seat.

o The aim is to eventually be able to sit down lowering yourself in control, with your back straight and your head up, without the use of your hands.

o   Do this everywhere, whether in your work chair, at the movie theater, on the bus, or at the dinner table.

iii) Squat

Standing up from a full squat will increase the benefit of standing. Equally lowering yourself down into a squat slowly will strengthen the back of your legs and stimulate blood pressure reflexes. If you travel to some countries that have more primitive restrooms, or you are caught out traveling through fields with no toilet in sight the ability to squat correctly, will come in handy.

If you have Type II diabetes, Dr. Michael Thorner, Head of Endocrinology, at the University of Virginia, School of Medicine, prescribes squatting three times every hour as the best activity treatment for his diabetic and pre-diabetic patients to keep their blood sugar under control.

Eventually, stand up from a squat unaided. That's what half the population in the world does every day.

Tip: *you squat every time you sit on the toilet. Become mindful of how you do it.*

iv) Stretch Up

Whether sitting or standing

- Stretch your arms up over your head as high as you can. Do this for 5-30 seconds depending on how it feels.

- You may waive your arms over your head or stretch them gently to one side or the other.

- Do this stretch every time you stand up or as many times as you feel like it in between.

- Likewise, if you spend a lot of time on your feet (e.g. if you work in a shop or restaurant), stretch up every 30 minutes.

v) Stretch Up

Need a mid-day break from sitting at a desk? Reach up against the pull of gravity whether you are sitting or standing. You can get a lot of benefits from stretching up when you need it most. Dr Peper a Psychology professor at San Francisco State University asks his students to stand up and wave their arms around after 30 minutes. He believes this Improves their cognitive skills and ability to learn.

(Debra Darnell 200 CYT-RYT, NY Strength Master Trainer)

vi) Inversion

You can be standing up or sitting, or upside down, as long as you are more or less aligned with gravity you benefit. You can hang on monkey bars at the playground or lie on the floor with your feet up leaning on a wall.

MOVE!

This is simply about any movement we would not think of as exercise. Increase it wherever and whenever possible.

Here are some suggestions and more are appended in the back.

- o Get up and visit your colleague's desk from time to time instead of sending electronic messages.
- o Park further away from your office, or the store.
- o Take the stairs up (and down) instead of using the elevator. Hold the railing as necessary for safety but aim to not use it to pull yourself up.

- Ride your bike to work, to the store or to see friends or attend events.
- Take your dog for a longer walk. Everyone will appreciate it.
- When you walk, make a point of keeping your head up your chest up and your shoulder blades pulled together. If you drop your head even a little, gravity will pull it down further making your spine bend which is dangerous to your health.
- Take public transport – the bus stop probably is not outside your front door, or even better, walk your child to school.

GET LIFE ACTIVE!

These are activities that allow you to accomplish things during the day, many of which you may look upon as drudgery. Instead, change your relationship to these actions, and you'll get more done and feel better at the same time.

- Put your clothes and shoes on and take them off at night standing up. Begin by leaning against the bed or standing by a wall for safety. You will quickly progress to better balance.

o Bend over to pick up something off the floor— a bit of fluff etc.

o Bend down to pick up a box – do not bend over but bend your knees to squat then come up with your back straight. If you cannot squat and come up with your back straight, do not pick up anything heavy this way until you can.

o Wash and clean your car regularly, inside and out.

o Clean the house and garage yourself; sweep, dust, and reach up and down.

o De-clutter. Start with your closet.

o Relearn to kneel. Whether it is to pray, to garden, or to scrub the floor it's the kneeling that matters.

> "While "cutting in" paint along a baseboard in a rather large living room I came to the realization that simply getting up from kneeling position, some 50 to a hundred times a day affected my overall health for the positive." Silvernails6, comment on Movement Heals on YouTube.

o Mix moving with social benefit. Call a friend to join you for a walk instead of lunch—alternate walking briskly and slowly.

- o Reach up to get something off the shelf. Step on a stool if you cannot reach it. Dust off cobwebs.
- o Stimulate your foot receptors by walking barefoot. Stand on your toes. Walk on your toes. Stand on your heels. Walk on your heels. If you must wear shoes use leather soles for better grounding.
- o Get up to get a drink of water; not only is the water good for you but you will need to get up and use the toilet more often.
- o Stand upright while queuing at the Post Office. Feel gravity pulling through you.

In the Life Science Division at NASA, I initiated a 10-minute stand.

- o Shake hands with people and say hello, how do you do? Make eye contact and say thank you.
- o Hold one arm at the wrist above the head and pull the other arm to one side — hold; then switch to the other side. Great first thing in the morning or do it in bed before getting up, or even while sitting in front of your computer.

b) Hold your arms in front of you - one arm stretched across against your body and pull the other at the wrist while bent at the elbow.

o Stretch up against gravity
o Raise your chest towards the ceiling or the sky. Your nipples should point forward not downward!
o Pull your elbows back till your shoulder blades touch.
o Relax: Lie down and think of gravity pulling you down through the floor.
o Relax– learn to float in water on your back without moving. This is Take public transport. Stand up holding onto a strap or pole — good for arm strengthening and balance.
o Carry the closest to being in space without gravity.
o Your own grocery bags to your car whenever you can. Great arm strengthening activity.
o Park as far as you can from your destination– supermarket, shopping, or work destination – less crowded parking – fewer dents and scratches on your car.
o Forego the ATM. Park and go inside the bank. Enjoy the social aspect.

PLAY

<u>It's a Myth</u> that because you are over 20 you cannot play anymore. Not true. Take every opportunity to play. Think back at what made you happy and what was fun. Maybe you cannot do everything but keep at it and enjoy. Go to the local playground to remind you what it was like. Or encourage your town to build you a playground for adults. Germany and Finland have them.

- o Swing on a swing in the park – great balance stimulus. Or at least use a rocking chair.
- o Hop, skip, jump; even little ones count.
- o Skip rope
- o Bounce on a trampoline (4G) – or a mini-trampoline.
- o Play on the see-saw with a friend.
- o Use the slide.
- o Walk on your toes. Walk on heels.
- o Spin 'carefully'. Aim eventually for three spins in one direction and then the other. Always go both ways.
- o Dance.

Florence is an elegant 92-years-old at Leisure World, in Silver Springs, MD who normally uses a walker. But at home, she turns on the music loud and dances to the beat pushing around her kitchen chair on rollers.

- o Ski– 85-year-old Ben in Wisconsin gets creative to get his skiing in.

- o Bounce on a trampoline. It's fun and generates 4G.
- o Take a short walk outdoors, fresh air (maybe). No cell phone. Concentrate on walking,

breathing, nature, and friends. Keep your head
and chest up –

Do not stoop.

> "Walk at moderate intensity or 100 steps per minute,
> hum or listen to the Bee Gee's song Stayin' Alive while
> you walk. The song's tempo is 100 beats per minute."
> Simon Marshall, PhD at UC San Diego

- o Play golf – carry your golf bags, walk
- o Play outdoor sports – tennis, bike ride, swim

5. Relaxation

When we are sitting are we relaxed? Not necessarily.
We can only be relaxed when we give in to gravity
totally. If we stand *still*, sit, or lie in bed to sleep or even
during sleep, some part of us is always moving and
contracting. Therefore, any muscle contraction begins
from an already contracted state. The nearest we come
to total relaxation is when we lie on the ground,
completely give in to gravity, and quieten our brains.
We can also feel total relaxation sometimes when we
float in water or are in space without gravity.
Relaxation allows a muscle to take full advantage of
each movement.

Free of Gravity

My friend astronaut and cardiologist Dr. Drew Gaffney told me that when he was launched into space in 1993 on the SLS-1 mission, as soon as the Shuttle separated from its boosters, he felt an immediate sensation of total freedom like he had never felt before or since on Earth, even though he was tightly strapped down to his seat.

Other Activities: make your own list

Home

- Squat to pick up a box instead of bending over – spares your back, and strengthens legs.
- Relearn to squat while waiting – strengthens legs, stimulates BP reflexes
- Squat, kneel, or sit on the floor to talk to or play with a child
- Kneel often if squatting is difficult –pray
- Stand up from a kneeling position until you can do it without leaning on anything

Enjoy sitting down to eat at the table. Structured sitting and social aspects of eating make this an important function.

- House clean. Sweep, vacuum, dust, mop floor, floors, clean bathrooms
- Make tea in a teapot. Practice pouring. Practice holding a full mug
- Carry your own grocery bags to your car
- Cook – Stir sauce in big pot on the stove
- Roll out pastry dough; knead bread
- Squat with awareness when sitting on a toilet
- Brush teeth with regular toothbrush manually
- Stand up during commercials when watching TV
- Stand up every 20 minutes when working at the Computer. Set a timer to remind you. It does not have to be exactly 20 minutes; anything between 20 or 30 minutes will do.

Spring/Summer

- Paint fence; Paint outdoors
- Garden – rake leaves, prune bushes and trees, bag-leaves, weed, Dig. Mow the lawn with a push mower
- Mulch
- Plant flowers and vegetables
- Sweep, wash driveway
- Clean out garage

- Tennis. Bend over to pick up the ball instead of bouncing the racket or helping the ball up with a racket
- Swim briskly
- Dance. Put music on loud and wave and shake your arms above your head.

Fall/Winter

- Rake leaves
- Sweep fluffy fresh snow on steps and driveway
- Shovel heavy snow
- Clean leaves from gutters
- Prune bushes
- Weed and mulch before winter temperature drops
- Play indoor tennis
- Sign up for Yoga or pilates classes
- Clear drains
- Paint indoors
- Play indoor tennis or Pickle Ball
- Take a walk whenever you can

Office/Work

Workouts are no antidote to death by desk job

- 04 July 2013 by **Richard A. Lovett**
- Magazine issue <u>2923</u>. **Editorial:** "<u>Don't take life sitting down</u>"

MICHAEL JENSEN is talking to me on the phone, but his voice is drowned out by what sounds like a vacuum cleaner. Or maybe it's a lawnmower. I'm used to bad connections, but Jensen isn't using Bluetooth on a busy freeway. He's in his office at one of the US's top medical research facilities.

"I'm sorry," he says when I ask about the noise. "I'm on a treadmill."

I'd had a similar experience earlier with David Dunstan, an Australian researcher who talked to me on his speakerphone as he walked around his office at the Baker IDI Heart and Diabetes Institute in Melbourne.

It's not that Jensen and Dunstan are hyperactive. Rather, both are exercise researchers looking into the link between sitting down and premature death. And what they have found is clearly disturbing enough for them both to make sure they spend ...

- Exhale deeply at traffic lights. It is healthy and calming.
- Park away from office building; avoids dents in car body

- Avoid elevators and escalators. Take the stairs instead.
- Pace while you speak on the phone. Do your phone calls standing or walking around your desk or up and down the hall. You no longer need to sit at your desk to take the calls.
- Drink lots of water, tea, and coffee throughout the day. It is healthy for your kidneys and you will have to get up to empty your bladder more frequently.
- Place your water bottle more than arm's length away so that you will need to get up to reach it. Or walk to the water fountain if you have one.
- Avoid texting, emailing, or calling fellow workers. Walk over to their workspace. they might stand up as well. Remain standing,
- Hold stand-up - meetings. They are more efficient and don't last as long.
- Place your printer, copier, and fax away from your desk
- Take the rollers off your chair. Makes you stand up to reach something.
- Replace your comfortable desk chair with a hard-backed chair –will encourage you to stand up more often

- Choose a chair with no armrests —encourage good posture, tone core muscles, and push elbows back every 15 minutes to relieve back and neck tension
- If you need it, get a sliding X-Finity desk or Stand-Desk computer terminal that allows you to slide up and down with your desk at intervals. Do your research and see which suits you best.

Sitting on a balance air pillow with small protrusion (such as Danskin's) – requires subtle balance and readjustment. It is good core training. It also helps elementary school boys in particular concentrate better.

Take Home Message

With the development of modern tools of convenience and entertainment has come a proliferation of mass sitting and with it a rapid, dangerous decline in the quality of our collective health. Even in many who exercise regularly.

Our grandparents moved all day without a second thought because they had to or felt the need to move. We no longer do that. We need to go back deliberately, mindfully at first, into the same habit-forming mentality until it becomes our way of life again.

The fantastic news is that it is possible to reverse these health impacts by being more like our ancestors: by using simple, frequent movements that challenge gravity and by doing so all day long, every day.

------

1. The fed's official Physical Activity Guidelines require 150 minutes of walking or 75 minutes of jogging per week, plus other exercises such as sit-ups. In general, the Centers for Disease Control says that 80% of Americans do not exercise enough. (Read More At Investor's Business Daily:

http://news.investors.com/politics-andrew-malcolm/050613-654907-gallup-poll-finds-americans-exercising-less-despite-michelle-obama-plans.htm#ixzz2SXpB9W9T)

[1] Levine JA. What are the risks of sitting too much? Mayo Clinic Q & A.2013.
http://www.mayoclinic.com/health/sitting/AN02082

[2] Duvivier BM, Schaper, NC, Bremers MA, van Crombrugge G, Menheere PP, Savelberg HH. Minimal-intensity physical activity (standing and walking) of longer duration improves insulin action and plasma lipids more than shorter periods of moderate to vigorous exercise (cycling) in sedentary subjects when energy expenditure is comparable.PLoSOne.(2013):8(2):e55542.doi

[3] Levine, JA, Schluessner,SJ, Jensen MD. Energy expenditure of non-exercise activity. Am J Clin Nutr 72 (6):1451-1454m 2000

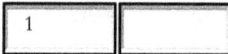

1	

Am J Prev Med. 2014 Jan;46(1):30-40. doi: 10.1016/j.amepre.2013.09.009.